Elvira Carver

How to Teach Geography

A Plan for an Elementary and a Scientific Course

Elvira Carver

How to Teach Geography
A Plan for an Elementary and a Scientific Course

ISBN/EAN: 9783337419868

Printed in Europe, USA, Canada, Australia, Japan

Cover: Foto ©Paul-Georg Meister /pixelio.de

More available books at **www.hansebooks.com**

HOW TO TEACH

GEOGRAPHY.

A Plan for an Elementary
and a Scientific Course.

By Elvira Carver,

*Teacher of Geography in the Westfield State
Normal School.*

REVISED EDITION.

BOSTON:
EDUCATIONAL PUBLISHING COMPANY,
1887.

Preface.

This little manual is prepared with the earnest hope that it may be of some service to those teachers who feel the necessity of giving oral instruction in geography, but are somewhat uncertain as to what and how much should be taught.

E. C.

Westfield Normal School, 1887.

INTRODUCTION.

Young children are eager to learn and have a wonderful aptitude for gaining knowledge from things ; for this reason they must be taught objectively. With them the teacher is everything, the text-book nothing.

The first years of a child's school life should be devoted to gaining that sort of knowledge and that sort of mental discipline which will enable him in the upper grades to come to new knowledge by thinking, rather than by reading what others have thought. For this reason this course in geography is divided into two, an Elementary and a Scientific Course. The first is adapted to the upper primary and intermediate grades, the second to the grammar and high school grades.

Much of the geographical knowledge required in the public schools can be gained more easily and more accuratel y from a map than from any de-

scription ; the early work therefore in this sub-
ject must be of a kind to enable the children
to interpret properly a map. They can be prepared to do this by studying the surface of their
own town, and in no other way. Moulds are in-
valuable, but a study of them must follow, not
precede, the study of the surface which they represent.

The children must then, as early as possible,
become familiar with their own region. They
should climb the highest points in their town, and
learn how these heights affect drainage and temperature ; should study their native streams, and
learn how they are formed, what their effect is upon
the surface, and in what ways such streams are useful
to man. The pupils should also be taught to represent by a scale the bodies of land and water
with which they have become familiar in this way.
After a time a map of a continent may be put into
their hands, and they may study it as they would
study that which it represents, for it will really be
to them an illustration of a continent. With their
previous training, their imaginations, as they look
at this map, will picture to them a great stretch of

country bordered by lofty mountains and traversed by mighty rivers, and from this map they can learn all they need to know about the physical features of the continent. Nor is this all; with the map and correct data from which to reason, a class can find out for themselves most of what is given in our school geographies. For instance, a child does not need to go to the printed matter of his book to learn that the Mississippi valley is a fertile section devoted to agriculture and capable of supporting a dense population. In studying climate he learned that the prevailing winds of this region are from the south-west; he sees they must take up moisture from the Gulf of Mexico; he knows they are moving to cooler latitudes, that by the lowering of their temperature the moisture will be condensed and fall as rain, and that therefore this part of North America must be fertile. In studying the divisions of land and water on the coast, he learned of New England's excellent harbors; in studying the mountains he discovered that in this region they approach the coast, that their slopes are short, and, that therefore, the rivers must be short and rapid, and so furnish water power. The causes that affect climate taught him about the cold current that comes down between our coast and the Gulf Stream, furnishing us with excellent fish. Having discovered all these things he sees that this populous section must be devoted to commerce, manufacturing and fishing, and he will be able to tell where the manufacturing and commercial centres are found. The object of the Elementary Course is so to train the children as to enable them, in the Scientific Course, to gain their knowledge in this way, instead of memorizing the printed matter in their text-books.

GEOGRAPHY.

ELEMENTARY COURSE.

This should begin when the children are about seven years of age, and should continue until they are ready for a scientific study of the earth.

See topics for a "Scientific Course in Geography."

OBJECT OF THESE LESSONS.

a. To give the children knowledge needed in the Scientific Course.

b. To train them to observe and think.

c. To train them to express their thoughts well, orally and in writing.

LESSONS FOR THE FIRST YEAR.

(Age about seven).

I. FORM.

Body

$a.$ ball $\begin{cases} \text{round.} \\ \text{long.} \\ \text{flat.} \end{cases}$ $b.$ surface $\begin{cases} \text{plane.} \\ \text{curved.} \end{cases}$

$c.$ line $\begin{cases} \text{straight.} \\ \text{curved.} \end{cases}$

II. SIZE.

$a.$ Exercises in estimating length by the eye, and with a ruler.

$b.$ Teach the table to 3 ft. $= 1$ yd.

III. DIRECTIONS.

North, south, east, west.

IV. SURFACE OF THE REGION.

$a.$ Take the children out occasionally to study some brook, river, hill or mountain in the town.

$b.$ Conversation with the children (using a mould) upon these things, and also upon anything they can see and are interested in as snow, rain.

SUGGESTIONS UPON FIRST YEAR'S WORK.

At the end of the year the children should be perfectly familiar with the forms, surfaces and lines named, ready to answer any question upon the direction or locality of objects known to them either inside or outside of the school-room, should be able to represent with considerable accuracy any length from one inch to three feet, to use the ruler readily in testing, should be acquainted with some of the natural features of the town and deeply interested in the things just about them. Drill upon the different topics must be carried through the year.

LESSONS FOR SECOND YEAR.

(Age about eight).

1. FORM.

a. *Surface.*
square, triangle,
circle, ellipse,

b. *Line.*
circumference,
diameter,
horizon,

II. SIZE.

Teach by actual measurement the table to 8 fur. $= 1$ m.

III. DIRECTION.

North-east, south-east, north-west, south-west.

IV. DRAWING BY SCALE.

a. Objects having length. (rulers, canes, fishpoles etc.)

b. Floor of the room and objects on the floor.

V.

a. Continue the study of the surface of the town by taking the children out;

MOUNTAINS,	RIVERS,
Slopes Valleys Plains, Fertile Lands, Sterile Lands,	Source, bed, branches, banks, etc. Flow — where water comes from — Stones and other materials carried by the river. Effect upon the surface. Uses { navigation. { water-power.

Study lakes and ponds of the vicinity in the same way.

b. Excite ideas of boundary lines, cities, villages, towns.

VI.

a. Make a mould of the town, representing a mile by an inch.

b. Use the mould to teach those natural features of the town which the children have not been able to reach, and to revive a knowledge of those visited.

VII.

Special lessons upon

a. The weather,

b. Plants and animals of the town which are good for food, shelter or clothing.

c. Important industries of the town, a lesson upon each.

d. The railroads.

e. With the aid of pictures and objects, give the children a little of the history of the town and some knowledge of its famous men.

VIII.

Topics the children may use in preparing their lessons, and in reciting,—

1. Draw a map — scale one mile by one inch.
2. Length and breadth of the town.
3. Boundaries.

4. Coast — oceans, seas, gulfs etc.
5. Interior — mountains, plains, rivers, lakes.
6. Weather.
7. Fertile lands, sterile lands.
8. Things that grow.
9. Useful animals.
10. Industries.
11. Railroads.
12. Villages, cities.
13. Interesting things.

SUGGESTIONS UPON THE WORK OF THE SECOND YEAR.

In giving the lesson upon the floor of the room remember that the object is not to teach the child to represent properly that surface, but is to excite in his mind ideas elementary to map drawing. It should not be given until the children are rapid and accurate in drawing lengths by their scale of one inch for one foot.

The picture when first drawn should be on a board on the north side of the room, and should be made by the pupils from their own observations and reasoning, and not by the teacher. Let them suppose the floor to be lifted up and represent one side after another, each one of the class, having of course, measured the sides for himself before the session began. If there is no board on the north side of your room, and no movable black board to be had in the building, obtain a piece of the black-board cloth sold so cheaply in all our large cities. If the suggestion made about drawing on the north board is followed, the children will mark properly north, south, east and west on the picture.

Before they begin to draw upon other boards, fix this one point in their minds, that east is at the right hand as one faces the picture. Say nothing about west, that will take care of itself. They will then be able to distinguish between east and west in using any map; a thing that a large number of pupils in the schools cannot do.

This lesson is given to train the children to in· terpret a map with reference to the location of the different parts of the surface represented, their direction from one another, and their relative sizes and positions. There would seem to be no special advantage in having the children spend a long time in learning to represent the desks, chairs and tables. One or two squares of pasteboard may be placed at different distances from the side of the room, and the children required to represent them in the right places on their diagrams of the floor. To do this, will necessitate hard thinking on their part, and will be a better preparation for map drawing, than to represent something the teacher has first represented for them.

If your town is a very small one, represent a mile by two inches, otherwise do not change the scale. During the entire year one inch should represent one unit, either a foot or a mile. The effort the children would make to master a new scale would destroy the impression you wish to produce by the drawing. For the same reason do not use fractions, unless your class are so drilled in them as not to have the least difficulty with them ; i. e., place your pasteboard two feet from the north and three from the east side of the room, not two feet and a quarter from the north and three feet and a third from the east side.

LESSONS FOR THE THIRD YEAR.

(Age about nine).

I.—FORM.

Other topics under space as degree, angle.

II.—SIZE.

a. Change the children's scale to one unit for ten.

b. Teach a square mile.

III.—SURFACE.

Continue the lessons upon wet and dry surfaces. Give the class new names for natural features as you find it necessary and convenient to do so. (The children should now begin to make excursions to interesting points outside their own town.)

IV.—OBJECT LESSONS.

a. *Weather, Clouds, Storms — rain, snow, thunder.* Have the children keep a record, from their own observations of the temperature, direction of the winds and the amount of rainfall during the year.

b. Plants and Animals.

The children should begin to study the conditions necessary to the growth of plants and those favorable to animal life. They should have lessons on the most important edible plants of the region and their relative value; also upon the best native woods and their uses.

c. People of the region.

Elementary lessons on races may be given.

1. *Races.* { Whites — German, Irish, French, why they come to our country. Blacks — A little of their story. (Inspire in the children love of country and sympathy for all races.)

2. *Government.*

a. Give the children some idea of the government of both town and state.

b. Tell them about our free schools, and the advantages they furnish poor boys and girls.

3. *Religion.*

Give them some idea of the people who worship many gods. Of people who wor-

ship one God as we do, but do not believe
in Christ. Explain the term Christians.

V.—MASSACHUSETTS.

*Teach the children from a mould. Topics for the
class to use in reciting on a state.*

I. Draw the map.
II. Size, — length, breadth.
III. Boundaries.
IV. Represent on' your maps and name :

Coast. — oceans, bays, sounds, islands, penin-
sulas, capes, isthmuses. Learn all you
can about these by studying your map,
then see if you can find something interesting about any of them in your books.

Interior. — mountains, slopes, plains, valleys,
lakes, rivers. Interesting things about any
of these.

V. *Weather.* — Find if the winds in all parts
of the state are like those of the town ; if
the sea-coast is warmer or colder than
other parts ; where the snow lasts longest.

VI. Represent the fertile lands, the sterile
lands ; tell what grows on each.

VII. Name any animals useful to man for food, shelter or clothing.

VIII. Tell what the people are doing in the different parts of the state.

IX. Represent the railways running through your own town, and show where they begin and end in the state.

X. *Towns.* -- Represent those on the railways, on the rivers, and tell what kind they are and why; also on the coast; a college town you know something about.

XI. *People.* — Are they all white; in what parts are the greatest number of foreigners found, why; are there free schools everywhere; who is the governor?

XII. *History.* -- In what year did the Mayflower come here; where did she land; whom did she bring; what did they come for; whom did they find here?

SUGGESTIONS FOR THE THIRD YEAR.
MASSACHUSETTS.

a. Excite an idea of a state, give the meaning of the term Massachusetts, show the state-seal.

b. As the children are now accustomed to represent ten miles by an inch, they can discover the length and breadth of the state for themselves, and also draw their map from the mould.

c. Train them to gain their knowledge as much as possible from the mould, that is, objectively. They are prepared to do this, because the mould of their town was the representation of a surface they had studied by making little excursions to all prominent points in it. Train them also to describe well what they see. Encourage originality of expression, but insist upon pure, clear, English.

d. Teach from the mould any natural features not found in the town, as peninsula, cape, gulf and bay. As new industries, unlike those of the town, are studied, give special object lessons upon them.

e. Appeal to the eye as much as possible.

The fertile sections may be represented upon the mould by powdered green crayon. A bit of granite and of wheat placed upon the mould to show where these are produced in the state, will interest the children and aid their memories.

f. Appeal to the reason as much as possible. Since the children have had proper lessons on the effect of slope upon rivers, and have studied the surface of Massachusetts, they can readily be made to understand why it is a manufacturing state.

g. Give the children a little of the history of their state. Story of the Indians, of the Mayflower; the date of settlement; what presidents Massachusetts has furnished the country.

h. With the aid of pictures, special lessons may be given upon interesting things, such as Bunker Hill Monument, Plymouth and Dighton Rocks, Hoosac Tunnel.

i. Associate the work of the reading class with that of the geography. If there is something in the reading book about Boston or Plymouth let that be the reading lesson the day the geography lesson is upon towns.

j. Now and then let the children commit to memory some fine bit of literature that relates to something in the geography lesson.

k. Have the children review, both by talking and by writing, the oral lessons you give them.

LESSONS FOR THE FOURTH YEAR.

(Age about ten).

I. — FORM.

Teach with the aid of a globe and other objects what the form of the earth is : that we live upon the outside ; and that we are kept upon the earth by attraction. (The last may be taught with the aid of a magnet.)

II. — SIZE.

Excite some idea of the great size of the earth by leading the children to find out how long it would take to travel around it in a boat, or by train, or on foot if such a thing were possible.

III. — USE OF GLOBE.

With the globe teach them how much of the surface of the earth is land, how much is water, and the position, relative sizes and names of the continents and the oceans etc. etc.

IV.—ORAL LESSONS ON SURFACE.

Continue the oral lessons upon dry and wet surfaces, but make the knowledge of the children more general, not confined to their own locality.

a. Dry surface.

Take up anything under highlands and lowlands that the class do not know, and that they need to know in order to study intelligently the surface of North America. (See topics for the children to use in studying a continent to determine what more must be taught your class, and topics under land, water and atmosphere in the Scientific Course for a plan.)

b. Teach more of the sources, direction and confluence of rivers,

c. Uses of lakes and rivers — drainage, navigation, water-power, irrigation.

d. Lessons that will give the children a more extended knowledge of the effects of water, especially of erosion, — ravines, canons, waterfalls, rapids.

e. Lessons upon special natural features, such as the Grand Canon of the Colorado, the Falls of Niagara.

V. — CLIMATE.

a. Require the children to make observations every day upon the direction of the wind, the forms of moisture, the temperature, and the effect of elevation, slope, and soil upon temperature.

b. Give them lessons at appropriate times upon these things and upon storms.

VI.—NORTH AMERICA AND THE UNITED STATES.

a. Teach the children from a mould of the country.

b. Have them recite from maps drawn by themselves.

Topics, the children in the intermediate grades, may use in reciting upon a continent, and upon an important political division like the United States.

I.   Form — draw the map.

II.   Size — length, breadth, tell how long it
        would take to ride across the country;
        to walk across it.

III.   Position — boundaries; if you have studied parallels show where the parallel of your town crosses the country.

IV.   Land and water of the coast.

  a. While drawing the map name and locate each important peninsula, cape, isthmus, sea, gulf, etc., as you represent.

  b. Represent and describe the islands of each ocean, and tell of any important sea, bay, or other body of water formed by them.

c. Tell anything interesting you have learned of the coast or of any body of land and water near it.

V.   Land and water inside the coast.

a. Find where the highlands are, the lowlands.

b. Represent the mountain ranges, giving their slopes and the highest point in each system.

c. Represent and describe each river system.

d. Interesting things about mountains and rivers.

VI.   Climate.

a. What the summer winds are, the winter winds, which bring rain and which snow.

b. Where snow is always found, where never.

c. Compare the climate of the coast and of the interior.

VII.   Productions and animals.

a. Tell where the fertile, sterile and mining sections are found.

b. Name the plants, animals and minerals of each section useful to man.

VIII.   Tell what the people are doing in these different sections, why?

IX.   How get from one part of the country to another.

a. What river systems may be used for this purpose?

b. Represent at least one railway that runs from ocean to ocean and (unless it runs through your town), represent also the railway that connects your town with the main line.

X. Towns.

a. Represent and tell something about the largest towns along this railroad, on each river system and on the coast.

b. Any other towns you are interested in.

c. How would you go by water from Chicago to New Orleans? Tell what goods, if a merchant, you would take with you and what bring back. Visit some places of intetest on the way and describe them.

d. Make up questions like " c " to ask the class, but be sure you can answer your own questions.

XI. People of different parts of the country.

a. Are they white?

b. Do they all believe in one God or are some of them pagans?

c. Is the ruler a king or a president? Tell his name and something about him.

XII. History.

When was the country discovered, by whom? (For other countries some of the questions must be changed a little.)

XIII.　Political Divisions.

a. Represent on your map the political divisions of North
    America, giving the boundaries and name of each.
b. Tell which of these divisions are independent, and
    which belong to some foreign power. Name the
    form of government of each.

Special study of the United States, using topics
like those for the study of a continent.

SUGGESTIONS.

a. When the children are ready to study the political
    divisions of the United States do not have them study
    each state separately, but in a group; i. e., let them
    study the New England States, the Middle Atlantic
    States etc.
b. Do not separate a map of a group altogether from the
    map of the whole. To do so, confuses the children
    with reference to the relation of a group to other
    groups and to the whole country. If, for want of
    space and time, you are obliged to allow the children
    to represent and study a group by itself, have on the
    board a large map of the United States, and during
    each recitation require one pupil to draw his map as
    a part of the large map, while the other members of
    the class are drawing the group simply.

LESSONS FOR THE FIFTH YEAR.

(Age about eleven).

I.

Lessons of the Fourth Year under I. and II. reviewed and extended.

II.

Continue the lessons given with the aid of the globe.

a. New terms defined.

1. Rotation — axis poles $\begin{cases} \text{north.} \\ \text{south.} \end{cases}$
2. Circles — large, small.
3. Large circles — equator meridian circle, meridian.
4. Small circles—parallels of latitude, tropics $\begin{cases} \text{capricorn.} \\ \text{cancer.} \end{cases}$

   polar circles $\begin{cases} \text{arctic.} \\ \text{antarctic.} \end{cases}$
5. Zones — torrid, temperate $\begin{cases} \text{north.} \\ \text{south.} \end{cases}$

   frigid $\begin{cases} \text{north.} \\ \text{south.} \end{cases}$
6. Latitude — north, south.
7. Longitude — east, west.
8. Hemispheres — north and south, east and west.

b. What makes day and night?

c. Day, night and twilight of the frigid zones.

d. What we shall see if we travel north, east, south and west.

e. Who live opposite us ; their day and night?

III.

Surface — new terms taught which the class are likely to come upon in studying.

The Continents.

1. Coast — archipelago.
2. Mountains — range, system, peak, pass, alp, ben, sierra.
3. Plateau.
4. Plains — prairie; savanna, steppes, selvas, pampas, desert, oasis.
5. Valley — river valley.
6. River — mouth $\begin{cases} \text{estuary.} \\ \text{firth.} \\ \text{fiord.} \\ \text{delta.} \end{cases}$

Cascade, rapids, waterfalls.

IV.

SPECIAL STUDY OF THE REMAINING CONTINENTS. (Suggestions).

FOR THE FOURTH AND FIFTH YEARS.

The children must learn the general topics — see topics for the study of the United States and of a continent in

Fourth Year. 1. Form; II. Size, III. Position, IV. Coast $\begin{cases} \text{water} \\ \text{land} \end{cases}$

V. Interior $\begin{cases} \text{land} \\ \text{water} \end{cases}$ and so on to XII. They will then be able to recite without being questioned.

The children have made a thorough study of their own town by studying it outside of the school-room, and so have experimental knowledge of all divisions of land and water within their reach.

They have studied their state, and have had lessons given them upon those forms of wet and dry surface not found in the town. Object lessons upon climate, and plant and animal life have been given them, and they have also learned to draw by a scale a surface with which they are familiar. The object lessons given the children have supplied them with terms and with the information they need to use in preparing their lessons. They are therefore ready to begin the work marked out for them in the Intermediate Geographies, which is largely that elementary study of the continents that precedes the child's entrance to the grammar school.

If required to do so by the school-board, begin to teach the earth as a whole by giving the children some idea of its form, size, motions, parallels, meridians, also of latitude and longitude. Such topics may be presented to a child, in a very simple and elementary way by the aid of balls, globes and other objects, still it is better to put them off as long as possible, because a young child's notions of these things are likely to be somewhat vague. If taught, difficult scientific definitions should be postponed until the Scientific Course.

Make everything as real as possible to the children. As soon as parallels of latitude and meridians are taught, begin to question the class daily concerning the circles passing through the town. Have a child tell whether in going home to his dinner he passed along a meridian or a parallel or neither, whether he walked towards the north pole or towards the equator. Let him see that every place on the earth is south of the north pole and north of the south pole ; that places on the equator have no latitude ; places on the prime meridian, no longitude, etc.

A small ball to represent the sun may be suspended above the globe. The children will see that one half the globe is in the light and one half in darkness. By rotating the globe you can make them understand how day and night are caused.

If a small ball be suspended as suggested you can, by elevating or depressiug the north pole of the globe, show the children where the direct rays of the sun are falling each day; and can also teach them that sometimes there is no day in the polar regions and at other times no night. Tell them how the sun appears to move when it does not go below the horizon. Also give them some idea of the long twilight of these polar regions and the advantage it is to the people.

Some of these lessons may be given upon the globe as if the children were travelling with you north, south, east and west from their own town. Travelling north with them into the Arctic regions, you may with the aid of pictures and specimens of all kinds, take up the appearance of the country, the climate, productions, animals, people, their mode of life, religion, dress, customs. Talk to the class about the children of the country, what they eat, wear, whether or not they are in school to-day. Taking the class east to the Atlantic ; you may teach something of its size, of the movements of its water and of the icebergs and animals found in it. With the aid of the globe the children can find what people live on the other side of the earth, and that it is night there when it is day here.

Lessons upon surface, climate and life should be continued by the teacher, objects being used when possible ; when not, pictures and drawings upon the board. Materials for such lessons may be found in any good geography. It does not require much skill to enlarge for the board the pictures found in the geographies, and to color them appropriately. Such pictures are of great value in teaching natural features, such as Niagara Falls, Grand Cañon of the Colorado, Yellowstone Park, Big Trees of California, Vesuvius. The following books may be helpful to you in preparing your oral lessons.

"Our World," No. II. Miss Hall, Ginn & Co.

"Eyes Right," Adam Stwin, D. Lothrop & Co. Boston.

"America Illustrated," Williams.

"Scribner's Geographical Reader" and "Primer."

"Elementary Lessons in Physical Geography," Geikie, Macmillan.

"The World at Home;" a Series of Geographical Readers, T. Nelson & Sons, London.

Special Lessons that may be given the children while studying the different continents.

NORTH AMERICA.

1. *Natural Features.*

   Niagara Falls, Yosemite, Yellowstone region, Grand Cañon of the Colorado, Big Trees of California, Mammoth Cave of Kentucky. Other interesting things such as the Panama Canal.

2. *Plants.*

   Rice, wheat, sugar, cotton, potato, tobacco, some woods.

3. *Animals.*

   Cow, horse, bees, cod, whale, oyster.

4. *Minerals, mines and quarries.*

   Gold, silver, copper, coal, iron, salt, granite, marble, etc.

SOUTH AMERICA.

1. *Natural features.*
   Amazon.

2. *Plants.*
   Coffee, chocolate, caoutchouc, woods.

3. *Animals.*
   Llama, ostrich.

4. *Minerals.*
   Diamond.

EUROPE.

1. *Natural features.*

   Lake Districts of England and Scotland, Fingal's Cave, Rhine, Alpine Region, Vesuvius; Pillars of Hercules.

2. *Plants.*
   Flax.

3. *Animals.*
   Reindeer, chamois, snail, sponge.

4. *Minerals.*
   Tin.

5. *Interesting things.*

The Alhambra in Spain.

The ruins of Italy and of Greece.

Pompeii and Herculaneum.

The Ear of Dionysius.

6. *Expressions to explain.*

Albania, Land o' cakes, Emerald Isle.

Sent to Coventry.

From Land's End to John O'Groats' house.

Kissed the Blarney Stone.

Escape Scylla and fall into Charybdis.

The crescent in Europe is waning before the cross.

ASIA.

1. *Natural features.*

Himalayas, Ganges, Indus.

2. *Plants.*

Palm tree, tea plant, bamboo.

3. *Animals.*

Elephant, camel, silk worm. cashmere goat, pearls and pearl fisheries.

Other interesting things.

Chinese Wall, Canton.

AFRICA.

1. *Natural features.*

Nile, deserts — mirage, sand storms.

2. *Other interesting things.*

Suez Canal, Pyramids.

KINDS OF INFORMATION.

Avoid crowding the minds of the children with unimportant facts. Children learn the names of too many capes, the lengths of too many rivers, the population of too many towns, and do not learn enough concerning a country's resources and those things which determine the location and the development of its people.

The children must have a general knowledge of a continent's physical features. This may be given them by requiring them, when upon a continent, to begin each day's recitation by drawing quickly upon the board a map, naming as they represent the most important divisions of the coast, and afterwards inserting the mountain and river systems. Seven or eight minutes are all the time needed for this part of the exercise. In this way the prominent physical features of the country will be forever fixed in the child's mind. Remember that the object of this map-drawing is not to train the child to draw beautiful maps, but to aid him in recitation and in retaining his knowledge.

SCIENTIFIC COURSE IN GEOGRAPHY.

A course for the grammar grades and for the lower classes in the high school.

The topics of this course are so arranged that, before the children take up the special study of each continent for the last time, they are made familiar with physical geography, a knowledge of which is necessary to an intelligent study of a continent, especially of its highest form of life, man. The earth is first studied as a whole by taking up its Form, Size and Motions; then its parts, the Land, Water and Air are studied. This work is followed by a special study of each continent.

Define Geography.

The Earth as a Whole. — Form, Size, Motions.

I. — FORM.

a. **Body.** **Solid.**

Spherical	Sphere
Hemispherical	Hemisphere
Spheroidal { prolate { oblate	Spheroid { prolate { oblate

b. Apparent form of the earth.

   1.   Show that hills and mountains have no effect upon the kind of surface.

   2.   The earth appears to have a plane surface.

c. Real form of the earth.

   1.   Show that when we look at a small part of a curved surface it looks plane, therefore we cannot tell by looking at it what the surface of the earth is but it has been found to be curved.

   2.   Not equally curved. The earth is a spheroidal body.

d. Show that a spherical body represents the earth.

e. Some simple proofs that the earth's surface is curved.

   1.   The way in which bodies moving over its surface disappear.

   2   Circumnavigation.

II. — SIZE.

a. Circles of the earth.　Circumferences.

　　Large　　　　　　Large

　　Small　　　　　　Small

Diameter.

b. Average circumference of the earth.

c. Average diameter of the earth.

d. Difference in length of the diameter from north to south and one of those from east to west.

III. — MOTIONS OF THE EARTH.

ROTATION.

a. Axis, rotation, poles, equator defined.

b. Special study of the axis of the earth.

　　1.　Relation to the direction of rotation.

　　2.　Relation to a person at the equator; at either pole; at a point between the equator and either pole; to a person in your town.

c. Rotation.

　　1.　Causes day and night.

　　2.　Causes the apparent daily motion of the heavenly bodies.

d. Poles.

　　1.　Show to what the north pole points and how to find the north star.

2. Place of the north star to a person at the equator, at either pole, at a point between the equator and either pole, to a person in the town.

POSITION OF A PLACE ON THE EARTH.

a. Latitude — north and south defined.

1. Parallels of latitude and their names.
2. How find latitude from a globe, from a map.
3. Comparative latitude.

> St. Petersburg, Mount St. Elias.
> London, James' Bay.
> Boston, Rome.
> New York, Constantinople.
> Richmond, Athens.

b. Longitude — east and west.

1. Meridian circle, meridian, prime meridian, longitude, defined.
2. Names of the meridians.
3. How find longitude from a globe, from a map.
4. Comparative longitude.
5. Degree of longitude.
   1. Length at the equator, at the poles, at sixty degrees north or south latitude. Nautical mile.

c. Explain the term antipodal points.

MOTION CONTINUED. — REVOLUTION.

a. Define revolution and orbit, and give the time and direction of the earth's revolution.

b. Tropics — Cancer, Capricorn.

c. Polar Circles — Arctic, Antarctic.

d. Zones — Torrid, Temperate, Frigid.

e. Seasons —Torrid $\begin{cases} \text{rainy,} \\ \text{dry.} \end{cases}$ Temp $\begin{cases} \text{spring,} \\ \text{summer,} \\ \text{autumn,} \\ \text{winter.} \end{cases}$ Frigid $\begin{cases} \text{sum'er,} \\ \text{winter.} \end{cases}$

Map of the whole earth, so that children may understand how meridians and parallels are represented. Represent some of the important meridians, as the prime, the 180th, the 90th, 45th, 135th west, upon a black globe; also the equator, tropics, polar circles; elevate the north pole and then represent the ball upon the board. Represent the eastern hemisphere in the same way.

Hemispheres —Eastern, Western; Northern, Southern; Land and Water.

THE PARTS OF THE EARTH.

Land, Water, Atmosphere.

I, — LAND.

Divide into Continents and Islands, by showing that a continent is a basin-shaped portion of land.

SPECIAL STUDY OF A CONTINENT.

Coast — peninsular, capes, promontories, isthmuses.

Relief — highlands, lowlands.

a. Highlands
- mountains, — range, system, peak, group, volcano ; where found.
- plateaus,
- valleys, — longitudinal, transverse.

b. Lowlands
- hills,
- plains,
- valleys.

c. Advantages of elevation.

    1. Effects drainage.

    2. Affects the distribution of moisture.

    3. Gives variety to climate.

II.—WATER.

Sea.

1. Description
- saltness, color, floor, depths, islands.
- temperature — equatorial waters.
  - polar waters — icebergs, ice fields, ice floes.

2. Divisions
- oceans.
- arms — seas—border, mediterranean — gulfs, bays. channels, straits, sounds.

3. Movements
- waves, cause.
- tides.
- currents — equatorial, polar, — direction of.

4. Surface Currents.

    a. North Atlantic Ocean.

    Warm Currents — Equatorial, Carribean, Mexican.

    Gulf Stream — temperature, shape, effect of spreading.

Divisions
{
A current by northern Europe.
Mediterranean current, return current.
}

SARGASSO SEA.

    Cold Currents — Greenland, Baffin Bay, Labrador, United States Current.

    b. South Atlantic.

    Warm Currents — Brazilian Current, Southern connecting current.

    Cold Currents — Antarctic Drift.

    c. Pacific Ocean.

    Antarctic Drift, Peruvian, southern-equatorial, northern-equatorial, counter-equatorial, Japan, Northern Pacific, Californian.

    d. Indian Ocean.

    Equatorial, Bay of Bengal, Arabian Sea, Mozambique, Return Current.

1. Compare the temperature of the eastern and western coast of North America.
2. The eastern coast of North America and the western coast of Europe.
3. The western coast of North America and the eastern coast of Asia.
4. The eastern and the western coast of South America.

Water of the Land.

Explain the Circulation of the Moisture of the Earth.

Moisture Received — Some sinks, some carried back to the ocean.

1. That which sinks.

Springs.

a. Classes of $\begin{cases} \text{constant, periodical.} \\ \text{cold, warm — hot, geysers.} \\ \text{Mineral } \begin{cases} \text{salt,} \\ \text{medicinal.} \end{cases} \\ \text{Explain causes of each.} \end{cases}$

b. Explain Wells — common, artesian.

c. Underground Rivers. $\begin{cases} \text{feed lakes.} \\ \text{flow back} \\ \text{to the sea.} \end{cases}$

Water carried back to the sea from the surface.

1. Rivers.

   a. Recall terms previously taught, as source, bed, mouth, and teach new terms such as channel, tributary, basin, water shed, river system.

   b. Formation — springs, lakes, snows, glaciers.

   c. Uses — drainage, navigation, irrigation, water power.

   d. Effect — Erosion $\begin{cases} \text{caused by current and débris.} \\ \text{forms ravines, cañons,} \\ \text{bottom lands.} \end{cases}$

   e. Deposition of debris causes deltas and sandbars.

2. Lakes.

   a. Recall terms such as outlet, inlet. head. foot and explain kinds, salt, fresh, and the cause of each.

   b. Source of Supply.

   c. Formation—Basins formed by $\begin{cases} \text{elevation} \\ \text{erosion} \\ \text{débris} \end{cases}$

   d. Uses — navigation, purify rivers, regulate the flow of rivers.

III. ATMOSPHERE.

a. Height, Composition, contains what?

b. Pressure.

   1. Due to what?

2. Amount at the sea level.

3. Effect of altitude.

4. Varies with temperature and moisture.

5. Thermometer, Isotherm.

6. Barometer. Isobar.

c. Moisture in the Atmosphere.

   1. Saturated Air, Dry Air.

   2. Condensation. Dew point.

   3. How may saturated air be changed to dry air.

   4. How may dry air be changed to saturated air.

   5. Forms of Moisture.

      Mist — fog, cloud $\begin{cases} \text{nimbus, cirrus,} \\ \text{cumulus, stratus.} \end{cases}$

      Rain.

      Snow·

      Hail.

d. Motion of the Air.

   1. Cause.

   2. Wind.

      Anemoscope.

      Kinds of winds — currents, storms.

I. *Special Study of Currents of Air.*

   a. System of — Ascending, Upper, Descending, Surface, Areas of low and high barometer.

b. Currents of each Zone.

Torrid Zone.

Equatorial calm belt or area of low barometer.

Upper or equatorial currents.

Cancer and Capricorn calms or areas of high barometer.

Trade Winds { Northeast.
{ Southeast.

North Temperate Zone.

Calm belts at the sixtieth parallels.

Southwest equatorial a return trade in summer.

Northeast polar wind in winter, deflected to north-west over eastern part of North America and Asia.

South Temperate Zone,

North-west equatorial in summer.

South-east polar in winter.

Frigid Zones.

Polar winds on the surface except about six weeks in summer.

c. Special study of the Equatorial calm belt.

1. Cause of change in position of.

2. Passes from fourteen degrees north to two degrees south.

3. Causes of Rain in the Equatorial calm belt.

4. Explain Rainy and Dry Seasons of the torrid zone. Where two seasons, where four.

I. *Some Local Winds of the Earth.*

  1. Monsoons — Indian Ocean, Gulf of Guinea, Mexico.

  2. Land and Sea Breezes.

  3. Etesian Winds.

  4. Desert Winds — Solano, Sirocco, Harmattan, Khamsin, Simoon, Föhn.

  5. Mountain Winds — Bora, Mistral, Puna, Pamperos, Northers of Texas.

II. *Special Study of Storms.*

  a. Kinds — rain, snow, hail, thunder, sand, cyclonic $\begin{cases} \text{hurricane.} \\ \text{typhoon} \end{cases}$

  b. Rain Storms.

    1. Occasioned by a current's ascending from the surface of the earth to the upper regions.

    2. Occasioned by a current's ascending highland.

    3. Occasioned by a current's moving into cooler latitudes.

4. Show that when a current descends into lowlands, or moves into warm latitudes it must become dry.
5. Rainfall of different zones.
    Rainfall decreases from the equatorial to the polar regions.
    Decreases from the coast to the interior.
    Amount of rainfall in the Old World, in the New World.
6. Explain the regions of greatest and least rainfall of each continent.

c. Snow storms.
  1. Occasioned how.
  2. Annual amount of snow.
  3. Temperature of the air at different heights The snow Line.
  4. Permanent snows — glaciers, avalanches.
  5. Uses of snow.

d. Hail-Storms.
  1. Hailstones — Form, size, structure.
  2. Cause of hail-storms.
  3. Geographical distribution of.
  4. Track of the storm.
  5. Destructiveness of.

e. Thunder storms.

    1. Cause of.

    2. Lightning $\begin{cases} \text{kinds of} \begin{cases} \text{chain, ball,} \\ \text{sheet, heat.} \end{cases} \\ \text{cause of} \end{cases}$

    3. Thunder — cause of.

    4. How determine the distance of a storm.

    5. Geographical Distribution of Thunderstorms.

    6. Destructiveness of.

    7. Use of lightning-rods.

f. Cyclones.

    1. Cause.

    2. Direction of wind about the calm centre.

    3. Rate of the wind about the calm centre.

    4. Area covered by the storm.

    5. Violence of the storm.

    6. Motion of.

        Path in the northern and in the southern hemisphere.

        Cause of the westerly motion of the storm in the torrid zone.

        Cause of the easterly motion in the temperate zones.

        Rate at which the storm moves.

Where are these storms most violent?

g. Storms of the United States.

   1. Summer Storms — Rain, Thunder, Hail.

   2. Winter Storms — Northeasters.

NOTE.—As rain, thunder and hail-storms have been thoroughly taught under the general topic, Storms, see pages 50 and 51, the teacher will have little to present relative to the Summer storms of the United States. From physical geographies and from such works as Prof. Loomis's Treatise on Meteorology, interesting facts concerning remarkable storms in our country, the average rainfall of different localities, etc., etc., may be obtained to give the class at this point.

NORTHEASTERS.

a. What are they?

b. Where do they originate?

c. Direction of the storm and rate of motion.

d. Changes in the barometer during the passage of the storm.

e. Veering of wind during the passage of the storm.
   When the western part of a storm passes over a place.

When the centre passes over a place.

When the eastern part passes over a place.

f. Rainfall in different parts of the storm.

g. Duration of the storm.

   3.   Signal Service of the United States.

a. Show how a storm may be predicted.

b. When was the Signal Service established ; at what stations ; for what purpose?

c. The service is under the control of what department of the government?

Explain the observations made, the instruments used and the signals given.

Use of the Signal Service.

CLIMATE.

Climate is affected by.

   a. Latitude.

   b. Elevation of the land.

   c. Slope of the land.

   d. Nearness to the Sea.

   e. Prevailing winds.

   f. Ocean currents.

Life depending upon Climate.

   I.   PLANT LIFE.

a. Show that the growth of the plant depends upon heat and moisture or upon climate.

b. Give a general description of the plants of the different zones.

c. Plants useful to man are commonly,

Trees — deciduous, evergreen, conifers,

Shrubs — deciduous, evergreen, conifers,

Herbs — grasses $\begin{cases} \text{meadow,} \\ \text{cereals,} \\ \text{bamboo,} \end{cases}$ pulse $\begin{cases} \text{peas,} \\ \text{beans,} \\ \text{lentils} \end{cases}$ edible $\begin{cases} \text{roots.} \\ \text{stems} \\ \text{buds.} \\ \text{leaves.} \end{cases}$

d. Give, with a diagram, the characteristic or representative plants of each zone.

e. Give, with a diagram, the particular plants of each zone.

f. Show that, on account of the lowering of the temperature of the land by elevation, there is a vertical distribution of plants corresponding to the horizontal distribution we have studied.

II. ANIMAL LIFE.

a. Show that animal life depends upon climate and vegetation.

b. Give a general description of the animals of the different zones.

c. Animals useful to man are wild and domestic.

d. Characteristic animals of each zone.

e. Particular animals of each zone.

f. Most important life of the earth is man.

III. HUMAN FAMILY.

a. Give the population of the world.

b. Races — Caucasian, Mongolian, Ethiopian, Malay, American.

c Give the location of the different races.

d. Special study of the Caucasian.

    1. Divisions—Aryan, Semitic, Hamitic.

    2. Show that this is a linguistic division.

e. Special study of the Aryan race.

    1. Divisions—Slavonic, Germanic, Keltic, Romanic races in Europe; Persians and Hindoos in Asia.

    2. Name the sub-divisions of the European divisions of the Aryan race, and tell where they are found.

Things that affect the development of a race.

I.—MODE OF LIFE.

Lessons on Occupations.

Agriculture, Grazing, Lumbering, Mining, Fishing, Manufacturing,

$$\text{Commerce} \begin{cases} \text{foreign,} \\ \text{domestic.} \end{cases}$$

$$\text{Means of Communication} \begin{cases} \text{railroads,} \\ \text{navigation.} \end{cases}$$

$$\text{Arts} \begin{cases} \text{industrial,} \\ \text{fine.} \end{cases}$$

II.—RELIGION.

a. Monotheistic.

$$\text{Christianity} \begin{cases} \text{Catholicism,} \\ \text{Protestantism,} \\ \text{Religion of the Greek Church.} \end{cases}$$

Judaism.

Mohammedanism.

b. Polytheistic or Pagan Religions.

Brahminism.

Buddhism.

Idolatry.

Feticism.

c. Confucianism.

d. Sacred Books of each monotheistic religion and of Brahminism and Buddhism.

e. Number and location of believers in the different religions.

f. With what race or races did the monotheistic religions originate?

g. With what race or races did the highest polytheistic religions originate?

Things that Affect Development continued.

III.—GOVERNMENT.

a. Republican,

b. Monarchi al.

Limited,

Absolute.

c. What are the best forms of government? Why?

d. What races have the highest forms?

e. What countries have the highest forms?

f. Had a republican form of government been tried by any people before our own republic was established; if so, with what success?

g. United States Government.

Departments—Legislative, Judicial, Executive.

Divisions of the Departments and Duties of each.

Legislative { Senate.
House of Representatives.

Judicial { Supreme Court of the U. S.
Inferior Courts of the U. S.

Executive $\begin{cases} \text{President.} \\ \text{Cabinet.} \end{cases}$

Give the names of the present cabinet.

Terms applied to races according to their state of development.

    a. Uncivilized—savage, barbarian, nomads.

    b. Civilized—civilized, enlightened.

    c. Semi-civilized.

Of the races studied, who are enlightened, etc?

Account for the condition of each race.

The pupils have now made a study of the form, size and motions of the earth; the configuration and relief of the land masses; atmospheric and oceanic movements; climate, plant and animal life, and especially man, including the distribution of races, and all conditions affecting this distribution; religion, government and whatever affects the civilization of races. They are, therefore, ready for a more intelligent and a more exhaustive study of each continent, than they were prepared for by age and by attainment when in the intermediate schools.

In studying a continent the class may take up those topics relating to the whole, those presenting the parts, and those which give a knowledge of the continent's fitness as a home for man.

The following topics for the class to use in . preparing their lessons, are made out in accordance with this plan.

CONTINENT.

TOPICS PRESENTING A CONTINENT AS A WHOLE.

a. *Form.*
b. *Introduction.*
c. *Size.*
> Greatest Length.
> Greatest breadth.
> Area.
d. *Position.*
> Latitude.
> Longitude.
> Boundaries.

TOPICS PRESENTING THE PARTS OF A CONTINENT.

1. *Outside and on the Coast.*

    a. Divisions of Water :

        Oceans.

        Seas, etc.

    b. Divisions of Land :

        Islands.

        Peninsulas.

        Capes.

        Isthmuses.

2.   *Inside the Coast.*

    a. Divisions of Land, or Highlands and Lowlands.

    b. Divisions of Water, or River Systems.

TOPICS RELATING TO THE LIFE UPON THE CONTINENT.

a. *Climate.*

Things that affect
- Latitude.
- Elevation of the Land.
- Slope of the Land.
- Nearness to the Sea.
- Prevailing Winds.
- Ocean Currents.

b. *Productions of Different Sections.*

Fertile and Sterile Sections—Productions.

c. *Animals Useful to Man.*

d. *Occupations Carried on in Different Sections.*

e. *Exports and Imports.*

f. *Great Thoroughfares.*

Water Communications. $\begin{cases} \text{Natural.} \\ \text{Artificial.} \end{cases}$

Railroads.

g. *Important Towns.*

Along these routes.

Manufacturing towns.

h. *People.*

Races.

Religion.

Government.

Advance in Civilization.

History.

i. *Political Divisions of the Continent.*